HOW TO GROW POTATOES

Beginners guide to growing, caring and harvesting potato

Larry Pat

www.ingramcontent.com/pod-product-compliance
Lightning Source LLC
Chambersburg PA
CBHW062311290526
45794CB00006B/2765

TABLE OF CONTENT

INTRODUCTION

Potatoes, the unassuming tubers that grace our tables in various forms, from crispy fries to velvety mashed goodness, have an illustrious journey that transcends mere culinary delight. Beyond their role as a staple in our diets, potatoes have etched a fascinating narrative through history, agriculture, and culture.

In this exploration, we peel back the layers of this versatile tuber to uncover its origins, its pivotal role in shaping civilizations, and its enduring impact on global agriculture and cuisine. Join us on a journey through the world of potatoes, where simplicity meets complexity, and where a seemingly humble vegetable reveals itself to be a powerhouse of sustenance, innovation, and cultural significance.Background

Potatoes (Solanum tuberosum) have a rich history dating back to pre-

Columbian times in the Andean region of South America. Native to Peru and Bolivia, potatoes were domesticated around 7,000 to 10,000 years ago by the indigenous people of the Andes. Initially, potatoes were cultivated for their hardiness and nutritional value, providing a reliable food source in the challenging mountainous terrain.

The potato's journey to global prominence began when Spanish explorers introduced it to Europe in the late 16th century. From there, the potato spread across the continent, becoming a staple crop due to its adaptability to various climates and soils.

Importance of Potatoes

The importance of potatoes extends beyond their role as a dietary staple. This subsection explores the multifaceted significance of potatoes in different aspects of human life.

1. Nutritional Value: Potatoes are a rich source of essential nutrients, including carbohydrates, fiber, and a variety of vitamins and minerals. Their nutritional profile makes them a valuable component of a balanced diet, providing energy and supporting overall health.

2. Economic Impact: Potatoes are a crucial commodity in the agricultural sector, contributing significantly to the global economy. The cultivation and trade of potatoes generate income for farmers, employment opportunities, and contribute to the overall stability of the food industry.

3. Food Security: Due to their high yield and adaptability to different climates, potatoes contribute to global food security. They serve as a reliable source of sustenance for populations around the world, particularly in regions where other crops may face challenges.

4. Culinary Versatility: Beyond their nutritional value, potatoes are celebrated for their culinary versatility. They can be prepared in numerous ways, from mashed and fried to baked and boiled, making them a favorite ingredient in diverse cuisines worldwide.

Chapter 1: Botanical Overview

Potatoes, belonging to the Solanaceae family and scientifically known as Solanum tuberosum, are not merely culinary staples; they are botanical wonders with intricate classifications and fascinating plant anatomy.

Classification

Kingdom: Plantae

Potatoes, like all plants, belong to the Plantae kingdom, emphasizing their status as living organisms capable of photosynthesis.

Division: Angiosperms

Within the plant kingdom, potatoes fall under the division of Angiosperms, highlighting their reproductive structures enclosed within flowers.

Class: Eudicots
Potatoes are classified as Eudicots, a group of flowering plants with two cotyledons in their seedlings, distinguishing them from monocots.

Order: Solanales
The Solanales order encompasses various flowering plants, and potatoes find their botanical companionship within this taxonomic order.

Family: Solanaceae
Within the Solanales order, potatoes belong to the Solanaceae family, also known as the nightshade family, which includes other well-known members like tomatoes and eggplants.

Genus: Solanum
The genus Solanum encompasses a diverse range of plants, and potatoes are a proud member of this genus, contributing to the botanical richness within Solanaceae.

Species: tuberosum

Solanum tuberosum is the specific species name for potatoes, signifying their unique botanical identity among other members of the Solanum genus.

Anatomy of a Potato Plant

Understanding the anatomy of a potato plant unveils the complexities that support the growth and development of the iconic tuber.

1. Roots: Potato plants have a robust root system that anchors them in the soil and facilitates the uptake of water and essential nutrients. These roots play a vital role in the overall stability and health of the plant.

2. Stems: The stems of a potato plant, characterized by nodes and internodes, support the growth of leaves and contribute to the transport of water, minerals, and sugars throughout the plant.

3. Leaves: Potato leaves are compound, consisting of leaflets attached to a central stem. The intricate network of veins within the leaves facilitates the exchange of gasses crucial for photosynthesis.

4. Flowers: Potato plants produce flowers with varying colors, ranging from white to lavender. These flowers are not only aesthetically pleasing but also serve a reproductive function, giving rise to the formation of tubers.

5. Tubers: The most distinctive feature of the potato plant is, undoubtedly, the tuber. Potatoes store nutrients in these underground structures, ensuring a reservoir of energy and sustenance for the plant during periods of dormancy or stress.

Chapter 2: Varieties of Potatoes

Potatoes come in a delightful array of varieties, each with its own unique characteristics, flavors, and culinary uses. In this section, we'll explore some of the prominent types, including Russet Potatoes, Red Potatoes, Fingerling Potatoes, and a glimpse into various other captivating varieties.

Russet Potatoes

1. Russet Burbank: A heavyweight in the potato world, is renowned for its high starch content and versatility. Its russeted skin and fluffy texture make it an ideal choice for baking, mashing, and frying.

2. Russet Norkotah boasts a smooth, tan skin and is favored for its excellent frying qualities. Its ability to achieve a golden, crispy exterior while maintaining a light, fluffy interior has made it a favorite for making French fries.

Red Potatoes

1. Red Bliss potatoes are known for their thin, red skin and waxy texture. These potatoes hold their shape well when cooked, making them perfect for salads, roasting, and boiling.

2. Norland Red potatoes are early-season red potatoes with a smooth texture and delicate flavor. It's versatility makes them suitable for various culinary uses.

Fingerling Potatoes

1. Russian Banana Fingerling potatoes are small and elongated with a buttery texture. Their unique shape and nutty flavor make them an excellent choice for roasting or adding to salads.

2. Purple Peruvian Fingerling potatoes boast a vibrant purple hue and a rich, earthy flavor. These visually striking fingerlings can elevate the visual appeal of any dish.

Other Varieties

While Russet, Red, and Fingerling potatoes stand out, numerous other varieties contribute to the potato's diverse world.

1. Yukon Gold potatoes are prized for their golden flesh and buttery taste. They work well in mashed potatoes, gratins, and a variety of other dishes.

2. All Blue potatoes, as the name suggests, have a striking blue-purple flesh. They add a colorful twist to dishes and are particularly popular for making vibrant mashed potatoes.

Chapter 3: Cultivation and Growth

Potatoes, being highly adaptable tubers, require specific conditions for optimal cultivation and growth. In this section, we will delve into the intricacies of meeting their soil and climate requirements, as well as the essential practices involved in planting and harvesting.

Soil and Climate Requirements

1. Soil Composition: Potatoes thrive in well-drained, loose, and sandy-loam soils. The ideal pH range for potato cultivation is between 5.0 and 6.5. This slightly acidic to neutral pH provides the potatoes with a favorable environment for nutrient absorption and root development.

2. Sunlight: Potatoes are sun-loving plants, requiring a minimum of six to eight hours of direct sunlight daily. Adequate sunlight promotes photosynthesis, leading to healthier plants and better tuber development.

3. Temperature: Potatoes are cool-season crops, preferring moderate temperatures between 60°F and 70°F (15°C to 21°C). While they can tolerate light frosts, extreme temperatures can adversely affect their growth.

4. Watering: Consistent and even watering is crucial for potato plants. Irregular watering can lead to issues such as cracking or misshapen tubers. Adequate moisture, especially during the tuber formation stage, is vital for a bountiful harvest.

Planting and Harvesting

1. Seed Preparation: Potatoes are typically grown from seed potatoes,* which are small tubers saved from the previous year's crop. Before planting, these seed potatoes should be cut into pieces, each containing at least one or two "eyes" or buds. This allows for multiple plants to sprout from a single seed potato.

2. Planting Depth and Spacing: Plant the seed potatoes in rows, burying them about 4 to 6 inches deep. Spacing is essential to ensure proper growth, typically ranging from 12 to 18 inches apart within rows, with rows spaced 2 to 3 feet apart.

3. Hilling: As the plants grow, It is common practice to hill soil around the base of the plants. This promotes additional tuber development and helps prevent the exposure of tubers to sunlight, which can cause them to turn green and produce solanine, a toxic compound.

4. Harvesting time depends on the potato variety and the intended use. New potatoes can be harvested about 2 to 3 weeks after flowering, while maincrop potatoes are typically ready for harvest when the tops of the plants naturally die back.

5. Harvesting Method: Carefully dig around the plants with a fork, being cautious not to damage the tubers. Harvested potatoes should be cured for a short period, allowing their skins to set, before storing them in a cool, dark place.

Chapter 4: Nutritional Value

Potatoes, often celebrated for their culinary versatility, also offer a rich array of nutrients that contribute to their significance as a dietary staple. This chapter delves into the nutritional value of potatoes, examining both macronutrients and micronutrients, and explores the various health benefits associated with their consumption.

Macronutrients

1. Carbohydrates: Potatoes are a primary source of complex carbohydrates. The starch content in potatoes provides a steady release of energy, making them an excellent choice for sustained fuel throughout the day.

2. Dietary Fiber: Potatoes are rich in dietary fiber, aiding in digestion and promoting a feeling of fullness.

The skin of the potato, in particular, contains a significant portion of this beneficial fiber.

3. Protein: While not as protein-dense as some other foods, potatoes do contribute a notable amount of this essential macronutrient. When combined with other protein sources, potatoes can be part of a well-balanced diet.

4. Minimal Fat: Potatoes are naturally low in fat, making them a healthy option for those conscious of their fat intake. However, the cooking methods and added toppings can influence the overall fat content of potato-based dishes.

Micronutrients

1. Vitamin C: Potatoes are a notable source of vitamin C, an antioxidant that supports the immune system, promotes skin health, and aids in the absorption of iron.

2. Vitamin B6: Vitamin B6 is abundant in potatoes, playing a crucial role in brain development, the production of neurotransmitters, and the metabolism of proteins.

3. Potassium: Potatoes are rich in potassium, a mineral that helps regulate blood pressure, balance fluids in the body, and support proper muscle and nerve function.

4. Iron: While not as high as some other sources, potatoes contain iron, contributing to the transport of oxygen in the blood and overall energy metabolism.

Health Benefits

The potassium content in potatoes supports heart health, helping to lower blood pressure and reduce the risk of cardiovascular diseases.

1. Digestive Health: The dietary fiber in potatoes promotes digestive health,* preventing constipation and supporting a healthy gut microbiome.

2. Nutrient Density: Potatoes offer a nutrient-dense option, providing essential vitamins and minerals with relatively few calories, making them a valuable addition to a balanced diet.

3. Antioxidant Properties: The presence of antioxidants, including vitamin C, helps combat oxidative stress in the body, potentially reducing the risk of chronic diseases.

Chapter 5: Potatoes in History

Potatoes have played a pivotal role in shaping human history, contributing to agricultural practices, population growth, and even facing challenges that left a lasting impact. In this section, we explore the historical significance of potatoes and delve into the unfortunate episodes of potato famines.

Historical Significance

1. Introduction to Cultivation: The history of potatoes traces back to the Andean region of South America, where indigenous people cultivated and domesticated them over thousands of years. Potatoes were a staple crop in the Inca Empire, and their cultivation techniques were passed down through generations.

2. Introduction to Europe: The Spanish explorer Gonzalo Jiménez de Quesada introduced potatoes to Europe after his expedition to South America in the late 16th century. Initially met with skepticism, potatoes gained popularity due to their adaptability to different climates and high nutritional value.

3. Population Impact: Potatoes played a crucial role in supporting population growth, particularly in regions where other staple crops faced challenges. The high yield per acre and nutritional content made potatoes a reliable food source, contributing to demographic shifts and urbanization.

Potato Famines

1. The Irish Potato Famine (1845-1852): One of the most infamous events in potato history is the Irish Potato Famine. A devastating outbreak of late blight, coupled with a heavy dependence on a single variety of potato (the Irish Lumper), led to widespread crop failures. The result was a catastrophic famine that caused the death and emigration of millions of Irish people.

2. Impact on Ireland: The failure of the potato crop had profound social, economic, and political consequences in Ireland. The population, heavily reliant on potatoes as a staple, faced widespread starvation and disease. The Irish Potato Famine is a tragic chapter in history, leaving a lasting impact on Irish culture and influencing patterns of immigration.

3. Global Implications: While the Irish Potato Famine is the most well-known, other regions faced similar challenges. Potato famines occurred in other parts of Europe, causing hardship and suffering. These events underscore the importance of crop diversity and sustainable agricultural practices.

Chapter 6: Delicious Potato Recipes

1. Classic Mashed Potatoes

Ingredients
- 4 large potatoes, peeled and diced
- 1/2 cup butter
- 1/2 cup milk
- Salt and pepper to taste

Instructions:

1. Boil the potatoes until tender, then drain.

2. Mash the potatoes and mix in butter and milk until creamy.

3. Add salt and pepper to taste.

2. Roasted Garlic Parmesan Potatoes

Ingredients:
- 2 lbs baby potatoes, halved
- 3 tablespoons olive oil
- 4 cloves garlic, minced
- 1/2 cup grated Parmesan cheese
- Salt and pepper to taste

Instructions:

1. Toss potatoes with olive oil, garlic, and Parmesan.

2. Roast in the oven at 400°F (200°C) for 30-35 minutes.

3. Season with salt and pepper before serving.

3. Loaded Baked Potatoes

Ingredients:
- 4 large baking potatoes
- 1 cup sour cream
- 1 cup shredded cheddar cheese
- 6 slices bacon, cooked and crumbled
- Chives for garnish

Instructions:
1. Bake potatoes until tender.

2. Cut a slit in each potato and fluff the insides.

3. Top with sour cream, cheddar cheese, bacon, and chives.

4. Potato Leek Soup

Ingredients:
- 4 leeks, sliced
- 4 potatoes, peeled and diced
- 6 cups vegetable broth
- 1 cup heavy cream
- Salt and pepper to taste

Instructions:
1. Sauté leeks until soft, then add potatoes and broth.
2. Simmer until potatoes are tender.
3. Blend until smooth, stir in cream, and season with salt and pepper.

5. Garlic Rosemary Roasted Potatoes

Ingredients:
- 2 lbs baby potatoes, halved
- 3 tablespoons olive oil
- 4 cloves garlic, minced
- 1 tablespoon fresh rosemary, chopped
- Salt and pepper to taste

Instructions:
1. Toss potatoes with olive oil, garlic, and rosemary.

2. Roast in the oven at 425°F (220°C) for 30-35 minutes.

3. Season with salt and pepper before serving.

6. Potato Gnocchi with Pesto

Ingredients:
- 2 lbs potatoes(Russet), boiled and mashed
- 2 cups all-purpose flour
- 1 egg
- Salt to taste
- Pesto sauce

Instructions:
1. Mix mashed potatoes, flour, egg, and salt to form a dough.

2. Roll into ropes, cut into gnocchi, and boil until they float.

3. Toss with pesto sauce before serving.

7. Hasselback Potatoes

Ingredients:
- 4 large potatoes
- 1/4 cup melted butter

- 2 cloves garlic, minced
- 1 tablespoon fresh thyme, chopped
- Salt and pepper to taste

Instructions:

1. Slice potatoes thinly, without cutting through.
2. Mix melted butter, garlic, and thyme.
3. Brush mixture over potatoes and bake at 400°F (200°C) for 50-60 minutes.

8. Potato and Cheddar Pierogi

Ingredients:
- 3 cups mashed potatoes
- 2 cups shredded cheddar cheese
- Pierogi dough
- Butter and onions for sautéing

Instructions:

1. Mix mashed potatoes and cheddar cheese.
2. Fill pierogi dough with the mixture, seal, and boil until they float.
3. Sauté in butter and onions before serving.

9. Potato and Corn Chowder

Ingredients:
- 4 potatoes, peeled and diced
- 1 cup frozen corn
- 1 onion, diced
- 4 cups chicken broth
- 1 cup milk
- 2 tablespoons flour
- Salt and pepper to taste

Instructions:
1. Sauté onions, add potatoes, corn, and chicken broth.
2. Whisk flour into milk and add to the pot.
3. Simmer until potatoes are tender, season with salt and pepper.

10. Broccoli and Cheese Potatoes

Ingredients:
- 4 large baking potatoes
- 1 cup broccoli, steamed and chopped
- 1 cup shredded cheddar cheese
- 1/2 cup sour cream
- Salt and pepper to taste

Instructions:

1. Bake potatoes until tender, scoop out the insides.

2. Mix potato insides with broccoli, cheese, and sour cream.

3. Refill potato skins and bake until cheese melts.

11. Sweet Potato Casserole

Ingredients:

- 4 cups mashed sweet potatoes
- 1 cup brown sugar
- 1/2 cup melted butter
- 2 eggs
- 1 teaspoon vanilla extract
- Topping: 1 cup chopped pecans, 1/2 cup brown sugar, 1/4 cup melted butter

Instructions:

1. Mix mashed sweet potatoes, brown sugar, melted butter, eggs, and vanilla.

2. Transfer to a baking dish.

3. Mix topping ingredients and sprinkle over the sweet potato mixture.

4. Bake at 350°F (175°C) for 30 minutes.

12. Potato and Egg Breakfast Skillet

Ingredients:
- 4 potatoes, diced
- 1 onion, diced
- 4 eggs
- 1 cup shredded cheddar cheese
- 2 tablespoons olive oil
- Salt and pepper to taste

Instructions:
1. Sauté potatoes and onions in olive oil until golden.
2. Make wells in the mixture, crack eggs into them, and sprinkle with cheese.
3. Bake until eggs are set, season with salt and pepper.

13. Potato Pancakes (Latkes)

Ingredients:
- 4 large potatoes, grated
- 1 onion, grated
- 2 eggs
- 1/4 cup flour
- Salt and pepper to taste
- Sour cream and applesauce for serving

Instructions:

1. Mix grated potatoes, onions, eggs, flour, salt, and pepper.

2. Form into pancakes and fry until golden.

3. Serve with sour cream and applesauce.

14. Potato Salad with Dill and Mustard

Ingredients:

- 5 cups boiled and diced potatoes
- 1/2 cup mayonnaise
- 2 tablespoons Dijon mustard
- 1/4 cup chopped fresh dill
- 1/2 cup chopped celery
- Salt and pepper to taste

Instructions:

1. Mix potatoes, mayonnaise, mustard, dill, and celery.

2. Season with salt and pepper.

3. Refrigerate before serving.

15. Scalloped Potatoes

Ingredients:
- 4 big potatoes, peeled and sliced
- 1/4 cup butter
- 1/4 cup all-purpose flour
- 2 cups milk
- 1 1/2 cups shredded cheddar cheese
- Salt and pepper to taste

Instructions:
1. Layer sliced potatoes in a baking dish.
2. In a saucepan, melt butter, whisk in flour, and add milk.
3. Stir in cheese until melted, season with salt and pepper.
4. Pour over potatoes and bake at 350°F (175°C) for 60 minutes.

Chapter 7: Challenges and Disease

Common Potato Diseases and Treatment

Potatoes, while resilient, are susceptible to various diseases that can impact yield and quality. Understanding and managing these diseases are crucial for successful potato farming. In this section, we'll explore some common potato diseases.

1. Late Blight (Phytophthora infestans)

Symptoms:
- Dark lesions on leaves
- White mold on undersides of leaves
- Browning and decay of tubers

Management:
- Use resistant potato varieties
- Fungicide applications
- Crop rotation and proper sanitation

2. Early Blight (Alternaria solani)
Symptoms:
- Dark lesions with concentric rings on leaves
- Premature defoliation
- Reduced tuber size and quality

Management:
- Crop rotation
- Fungicide applications
- Removal of infected plant material

3. Potato Virus Y (PVY)
Symptoms:
- Mottled or yellowing leaves
- Reduced tuber yield and quality
- Stunted plant growth

Management:
- Use certified disease-free seed potatoes
- Control aphid vectors
- Rogue infected plants

4. Early Dying (Verticillium spp.)

Symptoms:
- Yellowing and wilting of foliage
- Premature death of plants
- Dark streaks in vascular tissue

Management:
- Crop rotation
- Soil fumigation
- Resistant potato varieties

Pests and Challenges in Potato Farming and Solutions

In addition to diseases, potato farmers must contend with various pests and challenges that can impact the crop. Effective pest management strategies are essential for maintaining healthy potato plants.

1. Colorado Potato Beetle (Leptinotarsa decemlineata)
Symptoms:
- Defoliation of potato plants
- Larvae and adult beetles feeding on leaves
- Reduced yield

Management:
- Insecticide applications
- Crop rotation
- Biological control methods

2. Aphids
Symptoms:
- Yellowing and curling of leaves
- Transmission of viral diseases
- Sticky honeydew on leaves

Management:
- Beneficial insects (ladybugs, lacewings)
- Insecticidal soaps or oils
- Resistant potato varieties

3. Late Blight (Phytophthora infestans)
In addition to being a disease, late blight poses a significant challenge in terms of management due to its rapid spread and potential for causing severe crop losses.

Management:
- Regular monitoring and early detection
- Fungicide applications based on weather conditions

- Cultural practices to reduce humidity around plants

4. Soil-Borne Pathogens
Symptoms:
- Stunted growth and yellowing of plants
- Root rot and tuber decay
- Wilting, even with sufficient water

Management:
- Crop rotation
- Soil sterilization
- Use of disease-resistant potato varieties

CONCLUSION

In closing "The Potato Chronicles," we've uncovered the remarkable journey of the potato—its historical significance, botanical intricacies, nutritional value, and culinary versatility. From sustaining populations to facing challenges like the Irish Potato Famine, the potato's impact on history is undeniable.

Exploring its diverse varieties and nutritional benefits, we found the potato to be a resilient and adaptable ingredient. Our culinary expedition featured an array of recipes showcasing its versatility, while discussions on diseases and pests emphasized the importance of sustainable farming practices.

Beyond the farm, we delved into the cultural and culinary realms, discovering how potatoes influence traditional dishes and global cuisines. "The Potato Chronicles" is a celebration of this unassuming tuber's profound role in shaping our history, health, and culinary experiences.

As we conclude, may the love for potatoes continue to blossom in kitchens worldwide, connecting us to the rich tapestry of agriculture, culture, and sustenance. Happy farming, and may your potato adventures be as diverse and delightful as the stories we've uncovered together.